Railroads,

Biomass and

Synthetic oil

By

Robert H. Vollmerhausen

Table of Contents

Section 1

Railroads, energy and synthetic oil

Section 2

Expanded applications

Section 3

Putting it all together

Introduction

Railroads, Biomass and Synthetic oil starts with a basic premise: Whatever time we have left we must work to reduce our dependency on petroleum and also to reduce our demands on the environment.

Under the category of current events I accept the following basic conditions: We are leaving an era of cheap energy, plentiful land and resources and are entering an era of failing institutions and increasing human and environmental needs. Here we are. What do do?

The present work presents a set of engineering ideas, a tool kit suggesting that railroads and the associated right-of-ways might be utilized to provide increases in available energy and a means for recycling much of the refuse and trash that is now simply dumped into landfills.

Railroads, Biomass and Synthetic oil starts with a history of synthetic oil and describes a process by which railroads can be instrumental in growing substantial amounts of biomass as a means for sustainable synthetic fuel production.

The concept of developing a sustainable means for synthetic fuels production is more than a band-aid in our impending energy emergency. It represents a means for a limited, but sustainable continuance of industrial and technological transportation long after *Happy Motoring* has made its departure.

Railroads, Biomass and Synthetic oil is also about re-thinking how we organize and finance passenger train service. The fundamental premise is that in a 'down' economy' business entities that are not profitable on a 'stand-alone' basis might be profitable if paired with other compatible moneymaking activities. This work suggests developing a multi-faceted approach to energy and environmental problems:

- ❖ Passenger railroads are not profitable if operated solely by offering rail transportation;

- ❖ Passenger train service needs financial subsidy; and that subsidy is much less likely to be provided by the federal government in the years ahead.

- ❖ Likewise, recycling is not profitable when operated as a 'stand-alone' business activity, but recycling might be profitable if operated in combination with a coherent, large-scale synthetic fuels operation

This work describes a means for combining or consolidating both rail and 'recycling' into a composite business activity—a 'bundling' of operations into one all-encompassing business entity that, because of a synergetic unity, *might* be profit-making.

> The goal is to create a basket of business entities including rail and various recycling components that together produce a variety of energy products and transportation products that financially support each other.

Where rail alone is not profitable and where other environmental programs loose money, this work suggests that combining rail transportation with other activities can generate profit for both rail and for environmental protection. Railroads are the key, the backbone, to this concept.

One other goal of this work is to suggest ways to get the 'marketplace' actively behind environmentalism. Rather than have businesses externalize costs to the environment this work suggest strategies for broadening the scope of business activity to include environmental protection and beautification of the urban cityscape. How? By making those activities profitable.

Another theme of this work is to design rail transportation projects that are energy independent. That is, produce sufficient synthetic fuel to operate train sets independently of a common supply—while concurrently using railroad rights-of-way to produce large quantities of biomass feedstock for synthetic oil production. Large-scale biomass production supports synthetic fuels production, including diesel fuel, used by trains.

Redevelop passenger train service—and a means for *financing* the reestablishment of passenger train service by generating revenue from synthetic oil production.

Another goal is to better utilize transportation by rail and to concurrently develop other energy options to reduce our carbon footprint. We need to consider as many different types of *sustainable energy* as possible.

These concepts suggest ways to save energy—by actually using less energy (redeveloping passenger train service) and also for building *sustainable* infrastructures to assist in the production of synthetic liquid fuels.

Section 1

Railroads, energy and synthetic oil

Part 1

Synthetic Fuels

1

Synthetic fuel definitions and history

Wikipedia, the free Internet encyclopedia is the source for most of this part. Reference: www.wikipedia.org/wiki/Synethetic_fuel. It's an excellent article on both the history and technology of synthetic fuels. Their reference list prints out at two pages. It is an excellent article and recommended for any researcher interested in the potential of synthetic fuels to supplement petroleum.

According to *Wikipedia* 'Synthetic fuels are one of the few economically viable and industrially scalable alternatives to petroleum capable of providing a major source of the liquid transportation fuels required to run the economy and the only known non-petroleum source of aviation fuel.'

The *International Energy Agency* defines any liquid fuel processed from coal or natural gas to be synthetic fuel, although the term has several different meanings. Other organizations define synthetic fuel as any liquid fuel derived from coal, natural gas or biomass. The definition of synthetic fuels may also include fuels processed from oil sands and oil shale.

Processes may be either *direct* conversion—production of a liquid fuel directly from hydrotreating a feedstock to produce synthetic fuels—or *indirect* wherein the feedstock material is first converted into Syngas and the intermediate product is then processed into a liquid fuel.

Indirect conversion is a process where coal or natural gas is converted into a mix of hydrogen and carbon monoxide (Syngas) through a process of gasification or steam methane reforming and then the Syngas is transformed into a liquid transportation fuel. About 260,000 barrels of synthetic fuel is produced by the indirect method each day.

Synthetic fuels can be produced into diesel or high-grade aviation fuels. Germany, in World War II, produced a variety of synthetic fuels including special blends for aviation and for armored vehicles. The United States Air Force is developing its own program of synthetic fuels for its aircraft fleet.

A German chemist by the name of Bergius developed a process for the direct conversion of coal to liquids by hydrogenation. In this process coal is liquefied by mixing it with hydrogen gas and the system heated. After World War I several pants were built in Germany that were used during World War II to supply Germany with fuel and lubricants.

While a wide variety of projects are currently underway most utilize a method of mixing hydrogen and a feedstock under high temperatures and pressures to produce Syngas and then process the intermediate product into a variety of liquid fuel products. One example of a biofuel process is Hydrotreated Renewable Jet fuel.

This work does not discuss Peak Oil or the effects of a prolonged economic turndown due to resource depletion or the financial markets jumping the rails.

Rather the concentration here is that, for reasons best left for another time, we will need other sources of liquid fuels if any components of industrial civilization are to continue to function. I make the bold assumption that anybody reading this book already knows about Peak Oil.

2

Economics

The costs and economics of synthetic fuels varies on many factors such as feedstock used, the process and on the market. Analysts claim that synthetic fuels can be price competitive with oil down to about twenty dollars a barrel, but when other costs are added in such as carbon sequestration, or a change of feedstock to bituminous coal, synthetic fuels are then competitive with oil in the fifty-five to sixty dollar range.

Production of synthetic fuels must make money for investors. The ROI must be sufficient to induce potential investors to put money into synthetic fuel ventures—and indirectly to invest in railroads, because it is the railroads that provide the underlying foundation for large-scale biomass production, compost production and transportation and energy services. Still, everything depends on the economics of synthetic fuel production.

The National Energy Technology Laboratory, a division of the Department of Energy, conducted a study that found:

1.   Coal-to-Liquid plants equipped with carbon sequestration are competitive with crude prices as low as $86/barrel.

2.   Biomass-to-liquid plants are hindered by limited biomass availability which affects the maximum plant size that hereby limit potential economies of scale. This, when combined with relatively high biomass costs, results in diesel prices that are *twice* that of other configurations.

3.   The conclusion reached based on these findings was that the Coal to Liquid plants, with carbon controls implemented, offers the most pragmatic solutions to the nation's energy strategy dilemma. These options are economically viable when the price of petroleum is $86 to $95/barrel.

> Economies can change if plentiful low-cost biomass sources can be found. Developing a source of low cost biomass lowers the costs of inputs and improves profitability.

Moreover, there is a major consideration that the Wikipedia article doesn't mention—the United States is already producing about a billon tons of coal per year. That's a lot of coal. One of the difficulties with 'ramping up' synthetic fuel production based on using coal (CTL) plants is that the mining of coal in America, and throughout the world, is a well-established industry. As for, let's say, doubling coal production to meet the demand of a large-scale Coal-to-Liquid program the phrase 'easier said than done' comes to mind.

The United States has, depending on what numbers you believe, between 300 billion to 450 billion tons of proven coal reserves; enough for about one hundred fifty years at current usage. Here are the fundamental problems with coal:

1. While the rate of mining production (tons/yr) has increased about 1.2% per year since 1950 the heat content of coal varies widely depending on the type of coal mined. Hard coal with a high BTU thermal content is still mostly produced in the eastern United States.
   I do not know what type of coal was used in the NETL study, but we need to be cautious with 'estimated reserve' numbers, because a higher quality coal produces a higher quality liquid fuel product. (High quality coal = high thermal content.)

2. Even if high-quality coal reserves exist in sufficient quantity to sustain an aggressive CTL production mining millions of tons more coal presents a whole series of engineering (mining) and transportation problems.

3. Moreover, even if there are sufficient high-grade coal reserves it must be mined at a price that makes coal feedstock viable for conversion into liquid fuel.

Resource sustainability and the environment

Another reason to consider developing cheap and abundant biomass as a feedstock for synthetic fuel production is that to rely on coal, even if geologically and economically available, is to replace one exhaustible resource (petroleum) with another (coal).

What about oil shale?

According to: http://en.wikipedia.org/wiki/Oil_shale, an article entitled: *Environmental considerations,* 'Mining oil shale involves a number of environmental impacts... Oil-shale extraction can damage the biological and recreational value of land and the ecosystem in the mining area. Combustion and thermal processing generate waste material. In addition, the atmospheric emissions from oil shale processing and combustion include carbon dioxide, a greenhouse gas.

'Water concerns become particularly sensitive issues in arid regions, such as the western US and Israel's Negev Desert.'

This work advances the idea of using biomass in the production of synthetic fuels as opposed to incurring more water pollution and environmental damage processing synthetic fuel out of oil shale and oil sands.

The idea is to divert investment monies from environmentally degrading activities such as production of oil from oil sands to a cheaper, more sustainable and more profitable synthetic fuel alternative using biomass.

Biomass is the only *sustainable* method for producing synthetic oil. Production of biomass help to balance the production and sequestration of carbon dioxide while production of oil from shale sands is environmentally devastating

Cellulosic ethanol

Cellulosic ethanol is one type of biofuel produced from lignoclellulose—a structural material making up much of the mass of plants such as the stalks.

In 1898 the Germans developed an industrial process to hydrolyze cellulose to glucose and were able to produce around 50 gallons of fuel per ton of biomass. However, with the rapid development of enzyme technologies in the last two decades the acid hydrolysis process has gradually been replaced by enzymatic hydrolysis. There are two ways of producing cellulose ethanol:

❖ Hydrolysis on pretreated lignocellulosic materials using enzymes to break down complex cellulose into simple sugars followed by fermentation and distillation.

❖ Gasification that transforms lignoclellulose raw materials into gaseous carbon monoxide and hydrogen that are converted to ethanol by fermentation or chemical catalysis.

According to *Wikipedia* the U.S. could potentially produced 1.3 billion dry tons of Cellulosic biomass per year. That translates into 65% of American oil consumption.

Raw material is plentiful. Cellulose is present in every plant in the form of straw, grass and wood. Moreover, according to Wikipedia, even land marginal for agriculture could be planted with cellulose-producing crops like *switchgrass* resulting in enough production to substitute for all of the current oil imported into the United States.

In 2008 only a small quantity of switchgrass was used for ethanol production, because conventionally the land used is agricultural land—switchgrass competes with food production. A recent study by the University of Tennessee reported that as many as 100 million acres of cropland and pasture will be needed for switchgrass production to offset petroleum use by 25 percent.

Every 1,000 feet of track used in this concept, assuming a 50 foot cross section used for biomass production, would be the equivalent of one acre of crop land.

There are still approximately 233,000 miles of railroad rights-of-way left in the United States although that number decreases every day due to land 'development', abandonment and restricted access. Even twenty percent of the land area of those rights-of-way planted in switchgrass would provide *two million five hundred thousand acres* of growing area or enough area to produce approximately *seventy-five percent* of the oil now consumed every day in the United States.

This concept saves agricultural land for food production while simultaneously improving recycling, increasing the production of synthetic oil, co-generating electricity and improving railroad and synthetic oil industry profitability.

5

The potential and limitations of synthetic fuel

With aggressive development of synthetic fuels worldwide why worry about oil—or argue for the redevelopment of passenger train service?

The answer is twofold. First the world consumes approximately ninety million barrels of oil every day. Filling that many barrels with synthetic fuel is very unlikely. And secondly, in my unlettered opinion, most people—including most economists—don't really appreciate the effects of even a fairly small petroleum production shortfall. In 1973 oil supply was only about eight (8%) percent under demand and the economy was in a tailspin.

Although the Embargo was short lived, only about six months, the economy didn't really begin to recover until late 1975 or early 1976. An eight percent shortfall in oil supply caused real economic pain; and we are collectively using twice as much oil now then we did forty years ago.

In the 1970's there was talk of 'doing more with less,' and also 'less is more.' Over the last forty years we've just gone back to business as usual, but the next Big Trouble with petroleum will most likely be a long-term disaster for this country and for the world.

It might be that, if we are very lucky, we still have time to diversify our liquid fuels infrastructure with the objective of reducing our dependency on petroleum, but even the development of alternative transportation *and* multi-use infrastructure energy platforms, the diversification of transportation and energy, will only *soften* the effects of an oil crisis—We are just burning too much oil.

The age of Happy Motoring will come to an end sooner rather than later. We might, with railroads, maintain a semblance of industrial civilization, but only with a sustainable supply of fuel and with planning.

Part 2

Municipal refuse as Synoil feedstock

1

A brief history of New York City's Sanitation Department

Brendan Sexton was NYC's Sanitation Commissioner in the late 1980's. He knew they'd have to close the Fresh Kills Landfill early in the next century. It was closed in 2001.

A reporter asked Mr. Sexton 'What will we do with the garbage when you close Fresh Kills?' Mr. Sexton thought for a moment and substantially gave the following answer:

"The garbage is coming. We will try to burn it, but people will be unconvinced. We will try to put the garbage over here or over there, but people will meet us with their lawyers. But the garbage is coming."

Sexton went on to explain. "We will deal with the garbage, but we will deal with the garbage in a panic, in hysteria and at twenty times the cost and environmental damage, but yes we will deal with the garbage."

He was right. The garbage from Staten Island alone is 900 tons per day. It is sent by rail to a landfill in South Carolina.

Garbage from other boroughs is loaded onto 550 to 600 trucks and hauled to landfills as far as Wayne County, Michigan. As somebody once said of the U.S. government—this is the best money can buy. Can't we do better?

2

Biomass as the preferred feedstock source

One defining characteristic of synthetic fuels is that the same liquid fuel products are produced, often in the same plant, by different feedstocks: coal, biomass or natural gas; however, *only biomass* has the benefit of true sustainability.

Synthetic fuel plants can be designed to process coal and natural gas as well as biomass as feedstock. The process is flexible in that the proportions of one feedstock to another can change without adversely affecting the final product.

Synthetic oil production plants are processing biomass and coal-to-liquid fuels and are already planning on using significantly more biomass alongside coal. With a reliable source of biomass synthetic oil plants can make a transition from coal/biomass production to 100% sustainable biomass feedstock.

As *Wikipedia* says it, 'Some synthetic fuel processes can be converted to sustainable production practices more easily than others depending on the process equipment selected.'

This is an important design consideration as these facilities are planned and implemented, because additional room must be left in the plant layout to accommodate whatever future materials handling and gasification plant changes might be necessary to accommodate a future change in production profile.

Biomass is a sustainable feedstock for synthetic fuel production, but plastic waste is a close second so long as petroleum holds out. Part 3 discuses recycling plastics as a feedstock for synthetic oil production.

Synthetic fuels production is inherently flexible with respect to the feedstock used, however biomass has the benefit to true sustainability; and, it is carbon neutral.

Trains can haul and produce compost at the same time

Millions of tons of municipal waste are dumped into landfills, because we have chosen not to develop the technologies or the business models necessary to salvage this resource.

Railroads have a long-standing practice of bulk pricing; the more merchandise is moved in one shipment the lower is the unit price per ton. In this concept the compost and the plastic, used as feedstock for the production of synthetic fuels, has the advantage of bulk shipment pricing.

Railcars, designed and built for composting, can have self-contained water tanks and equipment for measuring and controlling the temperature of the cargo, rotating or turning it so as to evenly distribute the materials so that bulk materials are heated to at least one hundred seventy degrees Fahrenheit to kill bacteria or other pests.

Those railcars become paid-for cargo. The cargo is compost-in-the-making, or it will be, in three or four weeks as the train ships it to storage areas in a biomass growing areas. Part 4 starting on page 17 illustrates the design and operation of large-scale biomass operations.

The fact that such bulk material would have chemicals, such as paint thinner, or other chemical products not suitable for most agricultural applications probably would not be a problem so long as the biological product, biomass, is used for chemically producing synthetic oil—not food.

Any toxic waste brought up into the biomass plant would probably be destroyed in the Synoil production process, but even it wasn't it would be burned with the oil or otherwise captured in the manufacture of other products.

Things are good or bad by comparison. What is this hypothetical concept compared to? Consider: We are *already* hauling the waste! Except now we are dumping it into landfills and trying to forget about it until lawyers show up with a long list of environmental complaints.

Municipal compost as soil conditioner

Large land areas 'in the middle of nowhere' can be used to grow biomass—not food for humans or feed for livestock. The Southwest has miles of open empty desert along remote rail right-of ways. Of course *most* desert is not now used for any agricultural purpose, because such use is completely uneconomical—little or no water and land that is virtually sand with few nutrients and little capacity to hold water—Imperial Valley, California using Colorado River water is one exception.

Railroads and economics of biomass production:

❖ Rail can be used to ship bulk materials such as component-separated urban trash to storage bins at receiving stations within very large biomass growing sites.

❖ Railcars can be designed and built for the partial conversion of garbage and trash into a compost so as to improve soil conditions to the extent that the soil will support the growth of a biomass crop such as switchgrass.

❖ The economics of a biomass growing operation should be kept as cheap as possible, because the end result of growing the biomass is to convert it into synthetic fuels that must be priced to meet the market.

230,00 miles of rail right-of-way are a potential resource for growing large amounts of biomass for use as a feedstock in synthetic oil production. The concept is to use a combination of rail and passive and active energy technologies to grow large quantities of biomass.

This is a massive undertaking. Not to least of the problems is how to condition sand/ gravel and 'scrub' areas into land that will actually support mechanized and automated biomass production.

The key to improving agricultural production on any land is to condition the soil by improving water retention and the ability of soil to support microbial and plant life. *Composting* is one of the most basic and economical ways to improve soil. Any soil.

At this time most large cities and most counties haul their garbage to landfills. This is done for the usual reasons: money and politics. It's cheaper to throw garbage and most trash into landfills then separate the various components for reuse or recycling.

Most municipal waste consists of glass and plastic bottles and containers, paper, cardboard, table scraps and a wide variety of other items such as chemicals, both toxic and nontoxic. Production of biomass, strictly as input for synthetic fuel production, can benefit from using municipal waste to produce compost.

Recycling involves two major divisions:

❖ Compost production and

❖ Using recycled materials as a direct feedstock for synthetic oil production.

Part 3 is a discussion of the technology and economics of separating out these two basic components so as to isolate materials into the two material streams.

Part 3

Plastics as a resource

1

Plastics as a feedstock for synthetic oil production

Another sustainable feedstock for synthetic oil production is the abundance of plastics in urban waste. Our 'throw-away' society is throwing away another useable resource.

A common plastic bottle is illustrated below. Plastic bottles may be separated out from other components to facilitate an aggressive re-use and recycling program. Recycling more plastic is another means for increasing high-volume synthetic fuel production.

Plastic bottles with magnetic strips

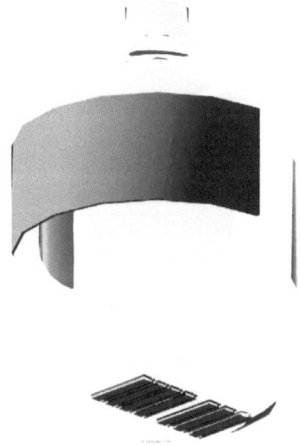

One method of separating various types of glass and plastic containers would be to affix magnetic strips to the underside of a bottle such as shown in the illustration on this page. This would not interfere with optical bar codes, but would provide one means of lifting specific bottles from the waste stream.

Another idea is to use *magnetic bar codes* that will not 'read' with optical scanners to further separate different plastic products. In the illustrations a conveyer belt (upper right) drops a variety of different plastic and glass containers down into a water tray.

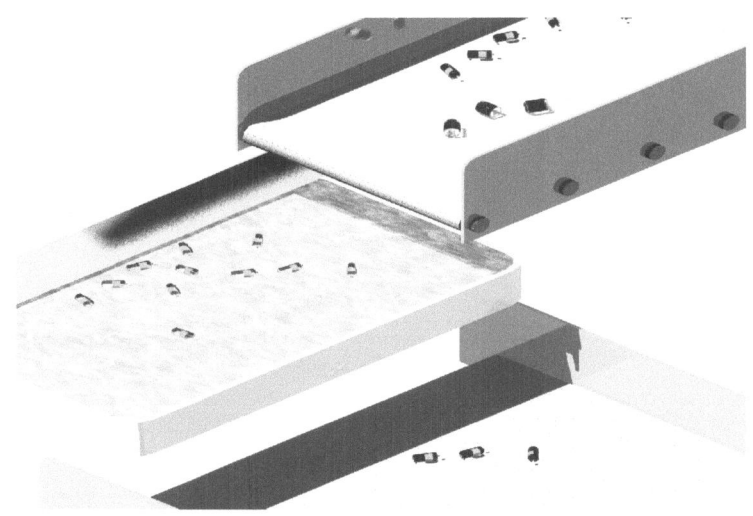

Prior to a plastics separation process magnetic separation has pulled metallic cans and bottle tops from the waste stream. Likewise paper, cardboard and polystyrene would have been separated out by air jets.

Separated paper products would be stockpiled for reuse or shipped for recycling—or, if not in condition for recycling then made part of compost production.

A combination of mechanical, water or magnetic processes can separate plastics of different types. The illustrations on this page and page 16 show how water can be used to separate out plastic bottles of different shapes.

The illustrations show a mechanical arm sweeping 'narrow neck' (floating) plastic containers off of the surface while the heavier glass and plastic (open mouth) jars are on the tray's bottom.

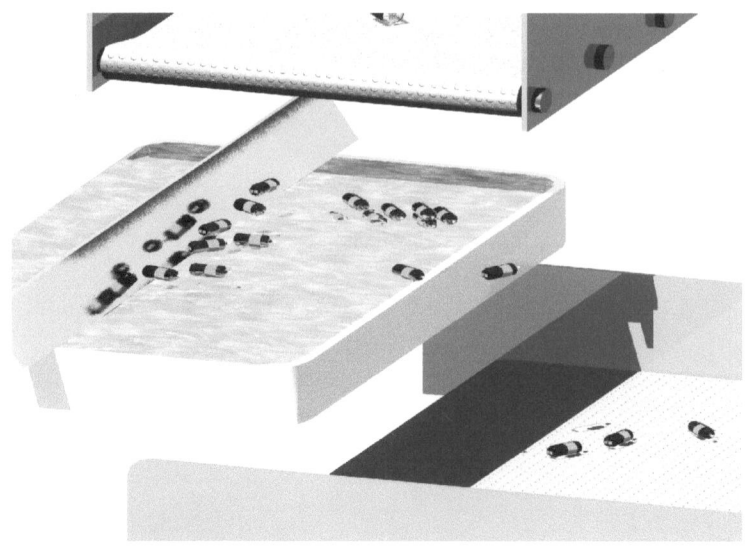

In actual practice this type of plastic separation would be a fast dynamic process. Plastic and glass containers would be separated out of the mix in a continuous process. These illustrations are merely to convey the idea that there are ways to improve our waste and refuse handling.

Reasons include:

1. To reduce pollution and dumping of materials in landfills and

2. Provide two or more commodity feedstock flows (biomass and plastics) for a large-scale production of synthetic oil and Syngas.

Part 4

Mechanics of biomass production

1

Human factors engineering

For illustrative clarity much of the biomass production and other associated operations are shown directly adjacent to the tracks. In practice, however, everything would be moved a distance away from the tracks.

Saltwater forests, for example, might involve growing woody plants. Trees, for example, that could interfere with the safe operation of a train—Sunlight filters through branches and throws a flickering pattern of shadows over the rails. This flickering is annoying at least and possibly causative of seizures in some people who don't even know they are susceptible to Photosensitive Epilepsy or flicker vertigo.

Also the area adjacent to the tracks must be kept clear for maintenance crews and equipment. A clean uncluttered view is also necessary for safely operating.

Different factors must be taken into account. The geography, human factors engineering, economic costs, environmental good—and the environmental costs—must be considered in planning such projects.

Not all railroad rights-of-way are created equal. Careful site analysis is necessary to determine what, if any, types of supplemental biomass or other economic activity might be developed on the site to make money for investors and the railroad. Provided irrigation water will be available one of the best, and cheapest, activities is to grow a biomass crop such as switchgrass.

2

Conditioning soil

Biomass is a crop. Maybe just grass, but a crop nonetheless. To get a good harvest all of the procedures developed for agricultural food production must be followed, except that in biomass production all of the associated factors must be extremely cost effective.

The concept of developing railroad bulk transport cars to haul and process separated urban refuse and trash into pre-compost material was previously presented.

The concept is to haul *and process* urban waste so that a pre-composted material is dumped into bins at a biomass-growing site. Trains are energy efficient so if the growing areas are near the rails we avoid the use of trucks to haul bulk compost.

It takes time and heat to turn table waste, paper and other materials into compost. The railroad cars pre-process those materials and deliver bulk materials to a biomass growing area.

At the growing site it is necessary to further that process along so that in a few weeks, with turning the materials and adding necessary water, those bulk materials become compost that, if tilled into loose earth such as sand, can condition soil to hold water and nutrients.

Once the site's growing areas have been suitably conditioned they can be planted with switchgrass or other biomass crop. While these areas are not used for food production they are producing a biological product that must be planted, irrigated, nourished and harvested economically.

Certain costs are more or less fixed such as the costs of running a railcar with bulk materials and dumping those materials into storage bins on site. Depending on topography, soils, climate and associated land costs these projects will potentially spread out over miles.

3

Automated machines

Biomass growing areas are located on property in proximity to a railroad right-of-way. Advantages to this arrangement:

❖ Railroad hopper cars can deliver bulk materials such as pre-processed compost and deposit bulk materials in storage bins alongside the tracks.

❖ Harvested biomass can be bulk shipped to processing plants for processing into a variety of synthetic fuel products. (Or, processed into synthetic oil on site.)

❖ Even assuming an automated biomass growing operation human supervision and maintenance is necessary—growing biomass adjacent to the tracks make transportation of personnel to and from remote sites fairly easy.

Lineal 'greenway' growing areas might stretch for miles. It would be too far for people to walk over and too remote to organize the work around people. With regard to the movement of personnel a small railroad siding can be built such that a 'railcar' (an automobile rigged to run on steel rails) can be parked off of the mainline.

The primary job of employees would be technical oversight of operations and mechanical repair and maintenance. Here again, if a piece of equipment needs major work, it can be transported to a shop by rail. Or, the shop comes to the site in the form of a railroad car.

The suggested biomass crop is switchgrass, because it is fast growing, disease resistant, grows in almost any soil and is easy to harvest—a mechanized lawn mower will do the job.

Long growing fields produce an agricultural product such as grass that is adapted for using automated machinery. (If it is technologically feasible to put a 'rover' on Mars it's possible to raise and harvest grass using automation.)

However, in the interest of economy and simplifying agricultural operations one machine should shoulder the work of many types of equipment. Rather than tractors, combines, harvesters and all of the equipment that is normally found on a working farm biomass production should involve only one or two machines.

The following functions are necessary:

* ❖ Bulk compost materials need to be turned and rotated.

* ❖ Hydroseeding. Grass seed is 'planted' by spraying a slurry of seed, water and liquid fertilizer over the site.

* ❖ Compost needs to be distributed from bins near the tracks to growing areas that might involve carrying bulk materials miles alongside the tracks.

* ❖ Harvested biomass needs to be transported to a rail transshipment area for loading onto railcars.

One or at most two different types of machines should do the job.

The following illustrations are an example of what an automated machine for use in biomass production might look like. It is designed to be simple to operate by remote control and simple to repair when broken.

A hollow box on wheels

It's fundamentally a hollow box with a plurality of individual augers arranged so as to 'pull' bulk materials in the front, store those materials inside the cargo box or eject them out of the back.

This type of machine 'tackles' bulk materials as shown in the illustration below. Bulk compost needs to be turned and redistributed periodically and wetted down to further the composting process.

When railway hopper cars dump pre-processed materials into bins the same train also delivers bulk water to be stored in bins on site.

This water is sufficient for maintaining the composting operation, but not nearly sufficient to irrigate the biomass crop. The water tanks on the sides of the machine are used for composting and for hydroseeding operations.

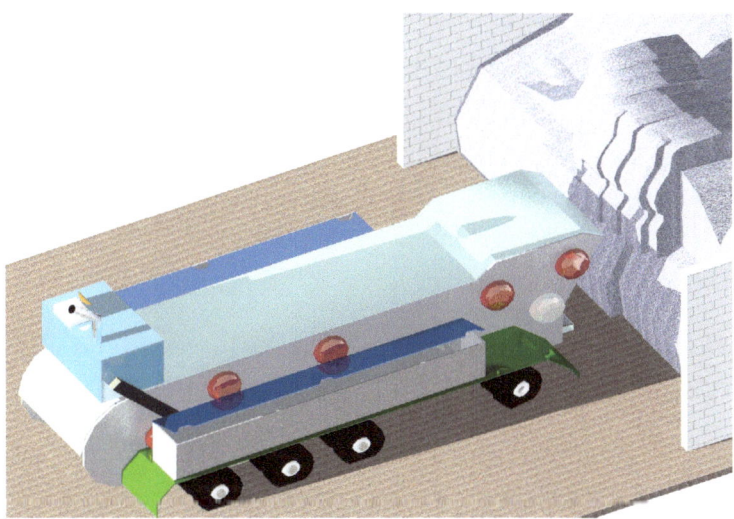

Water boxes sit on the fenders and provide water to compost or other bulk materials. Although the water boxes also have sprinkler heads for irrigating they are only used for specific areas such as after the machine has spread compost or other soil conditioners.

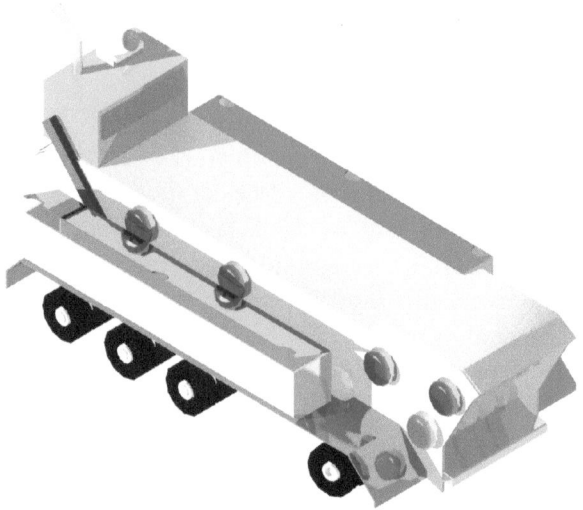

A diesel engine sits on top of the cargo box and the power train (in black slanting down the sides) provides power to the wheels.

It's a machine built for direct 'throughput' of materials. Individually controllable augurs 'bite' into bulk materials and pull material into the cargo box.

The augurs on the front of the machine swing up or down depending on the application. When harvesting is needed the steel blades act as a sickle to cut the grass and 'throw' it back into the cargo box.

Internally, augurs powered by individual electric motors move bulk materials from the front to the back of the cargo box.

On the front and back of the machine augurs do double duty of distributing and spreading bulk materials on the site as well as using the front to cut and 'throw' grass into the cargo compartment while the augurs on the back work bulk materials into the soil. Water boxes wet the soil to complete the operation. The following illustration shows the back augurs disposed to working compost into the soil.

A machine designed along the lines of the one illustrated could be used to transport and work composted materials into the soil, to 'hydroseed' areas, to cut or mow the grass when ready for harvesting and do other necessary work such as 'turn' bulk materials in the composting process.

In operation they would further process bulk materials brought to the site by train into compost and would deliver it to the growing areas.

Likewise, the same machines would till the compost into the earth and spread and irrigate materials along with planting (hydroseeding) the crop.

At harvest the machine mows the crop and 'throws' the harvest into the cargo bay for delivery to bulk storage bins where it is dried and possibly bailed for delivery to a synthetic oil production site—or depending on the operation the site may have a modular synthetic fuel production facility. It's more economical to ship liquid fuel than to ship switchgrass in bulk.

At full size these machines, as illustrated, would be about seven feet wide and ten feet high and would therefore be easily transported by rail from one operational area to another.

Part 5

Mechanics of biomass irrigation

1

Background

This part discusses irrigating biomass. Most of the areas for growing biomass would be located in the Southwest or other areas that may be subject to draught conditions. And, in many cases these biomass-growing areas would be miles from any locality.

Our impending dilemma with resources doesn't end with petroleum, but extends also to water resources. Some projects for growing biomass as feedstock for synthetic fuel production might be sited in areas of the far Southwest that are now experiencing an extended drought.

This work focuses on large-scale projects of growing biomass and on aggressive recycling of urban waste with the objective of producing large quantities of synthetic liquid fuels.

There are many business challenges with growing biomass, but primarily it is the cost. Ways must be found to minimize economic costs while maximizing the quantities of biomass grown and the subsequent quantities of synthetic fuels produced.

Conditioning the soil is one challenge. Providing sufficient irrigation water to the biomass crop without spending money on PVC pipe and conventional sprinkler systems is another challenge. Again, overall costs must be kept down.

Yet another challenge is to irrigate biomass crops without further depletion of geologic water supplies and without competing with local ranchers for a limited and often scarce resource—irrigation water must be found in other ways. This part looks at ways to irrigate large-scale biomass without the conventional use of drilling wells.

Conventional agriculture follows the lead of our overuse of oil: We drill for water. The aquifers in many parts of the United States are facing the same geological over-depletion as petroleum. And yet, rather than turn to other technologies, such as using fog fences, we prefer to pull resources out of the ground. A suggestion is made that other non-traditional sources of water should be considered.

The mechanics of condensing water from air are well understood, but the problem has always been the cost of the technology and the associated cost of energy to drive the technology.

The next segment discusses fog fences and trees as passive means for capturing water from air. The following segments also look at a way of using piezoelectric modules to generate electricity by salvaging mechanical energy as a train rolls over a set of piezoelectric crystals.

This 'active' technology has the potential to bring a new, more efficient, means for wringing moisture out of air due to a drop in temperature of ambient air so that water vapor condenses in greater volume.

This technology uses piezoelectric devices to harvest otherwise wasted mechanical and thermal energy as a train rolls over a PE device. The electrical energy produced by the piezoelectric device is converted to AC and then used to run air pumps that put high-pressure air into tanks. The following segments illustrate and give more detail on how this 'rain-making' methodology works.

2

Fog fences

Fog contains from one-half gram to three grams of water per cubic meter. Since the weight of water in fog is very low the 'settling' rate is also very low—fog simply goes where the wind takes it.

Over the last hundred years, however, various devices have been developed to condense fog droplets into water and one of the most economical and simple of these devices is the fog fence.

Since wind will take fog around a solid barrier the fog fence is simply a vertical structure with a mesh material strung between uprights.

Fog fences are easy to build and relatively passive; that is, they don't require any constant care or maintenance. They should be placed on the crest of a hill or other topology so as to catch the prevailing wind.

Fog fences have been used with mixed success in many countries from South Africa to Latin America. They do work best, however, in a political environment where the local population knows that the fences are in place to support agriculture and (usually) not to provide reliable drinking water for humans.

In this application fog fences are used as part of a biomass growing operation so that they is no confusion about the utility of the fences—they are not used to provide water to humans or livestock, because fog fences do not have the reliability we have come to expect from our water delivery systems.

3

Trees

Water is a scarce commodity especially in the Southwest where biomass-growing operations might be built if a sufficient source of water can be found.

Each potential site would need to be monitored carefully and analysis made of the potential for capturing fog and other sources of irrigation water without infringing on local agriculture or ranching operations.

According to: www.rexresearch.com/airwells.htm the Roman author Pliny the Elder mentioned the Holy Fountain Tree growing on the island of El Hierro in the Canary Islands. "For thousands of years…the people there obtained most of their water from the trees, the leaves of which captured fog.

Japanese Black Pines and other species of trees with needles are especially adept at slowing the velocity of a prevailing wind and capturing moisture as fog condenses, especially in early morning hours, on cold foliage.

Fog fences are made up of a fine mesh that does not impede the flow of air, but merely slows the air down to where tiny droplets are condensed and the moisture drains into pipes for collection.

In areas designated suitable for the growing of biomass, such as switchgrass, trees may be used as a substitute for fog fences or used in combination with fog fences. The dense bramble of needles making up a stand of pine, for example, has the same mechanical effect of slowing moisture-heavy air and allowing the deposition of moisture on the leaves.

The moisture yield of trees or fog fences is limited by local conditions such as climate and topography. Both trees and fences should be placed high enough on the crest of the terrain to capture as much of the prevailing wind as possible.

Trees also provide pine mulch and other materials for composting and soil conditioning. The primary reason for trees in the design of a biomass growing area, however, is to assist and complement fog fences in harvesting water from air.

In the current concept trees should be placed away from a railroad right-of-way and to the north of the tracks—in the northern hemisphere the sun follows an arc in the southern sky so that shadows fall to the north.

The construction of large-scale biomass growing operations should be done so as to leave the right-of-way free of visual and mechanical clutter.

As seen in the illustrations starting on page 33 biomass growing areas are lineal in basic shape. They might extend for miles depending on land, economics, soil and climate.

In this biomass production concept trees not only harvest water from fog and dew, but also sequester carbon and provide a source of revenue from the sale of trees to municipalities and landscapers.

There is also one other 'active' method of capturing water from air. That method is to use piezoelectric technology to harvest otherwise wasted mechanical and thermal energy as train roll over the tracks—that captured energy can be used to augment and increase the efficiency of capturing water from air. A discussion of piezoelectric technology starts in segment 4.

4

Salvaging wasted mechanical energy

Israelis have developed a test roadbed that uses the weight of automotive vehicles to compress/stress piezoelectric crystal modules embedded in the roadway to generate electrical current.

Innowattech is a research and development organization with facilities in the *Technion Institute, Haifa, Israel. Innowattech* specializes in development of custom piezoelectric crystal module generators.

The *Innowattech Piezo Electric Generators* (IPEG ™) harvests parasitic mechanical energy from roadways, *railways*, airport runways and even pedestrian walkways.

I. P. E. Generators are modules embedded under a railway track, illustrated on the top of page 30, and convert otherwise wasted mechanical and wasted heat energy (as the weight of a train deforms the rail) into useable electricity.

This technology does not actually produce energy, but it does salvage mechanical energy that is otherwise wasted as the train's weight deforms the rails. Such salvaged mechanical energy can be put to good use through the use of compressed air.

This technology captures otherwise wasted energy as trains roll over the piezoelectric modules. That harvested energy can be converted into electricity for pressurizing air tanks.

Compressed air is another way of accelerating the condensation of water from air. This technique is described in the next few pages. It works on the principle that cooling air reduces the dew point temperature so that as air cools water vapor condenses into liquid water.

By converting mechanical stress into electricity the idea is to use that electricity to pressurize air into air tanks that are subsequently used to 'spill' cold air into a prevailing wind.

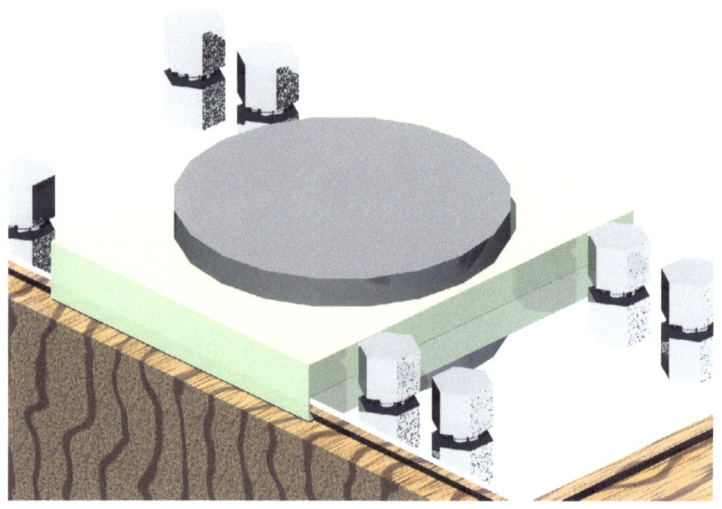

Electrical output in a rail infrastructure equipped with the *Innowattech* system is said to salvage 120 kW per hour from a one km stretch of railway or subway. As a train's weight deforms the rails piezoelectric crystals convert mechanical stress into electrical energy. Electricity is stored in a trackside apparatus (in blue) as shown in the next illustration.

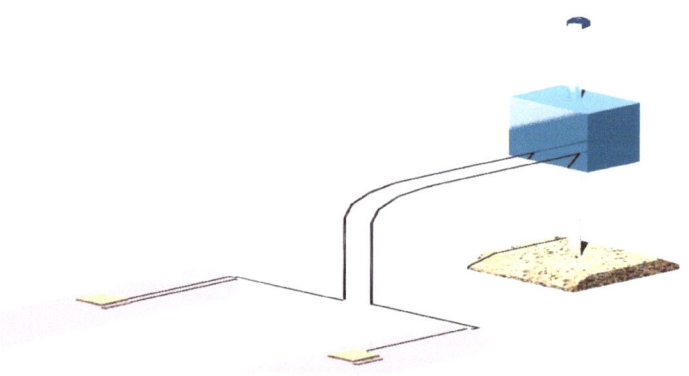

Direct electrical current is produced and is then converted into alternating current so as to make it useable throughout an electrical network.

5

Piezoelectric energy harvesting and irrigation

The previous part described piezoelectric technology and railroads to harvest otherwise wasted mechanical and thermal energy. This method of converting mechanical energy into electricity can be used to operate such things as railroad crossing gates in remote areas.

The advantage of using technologies such as piezoelectric to harvest mechanical energy is that it can be used in remote areas without running long-distance electrical wires or other utilities. Piezoelectric can also be used to provide a source of electrical energy for other applications.

Compressed air has many uses

According to *'A sourcebook for Industry, Introduction to Industrial Compressed Air Systems,'* "distributing energy through compressed air is more energy efficient and air-driven power tools tend to be smaller, lighter, and more maneuverable than electric motor-driven tools. They also deliver smooth power and are not damaged by overloading."

1.   Stores energy or,

2.   Provides repair shops, maintenance centers and manufacturing facilities (associated with the railroad right-of-way) with high-pressure air to power air tools, heat and ventilate buildings and co-generate electricity.

According to W.A. Amos, 1998 *'Costs of storing and Transporting Hydrogen'* (NREL/TP-570-2505) 'Thick, massive salt deposits could be used for compressed air energy storage (CAES) and hydrogen storage. There exist CAES plants in salt in Huntorf, Germany, for example, where a 290 MWe plant with two 10 million cubic feet caverns is located.

Compressed air provides flexible output

Compressed air is an economical way to store energy and it also provides a way to convert stored energy into a variety of useful end products: Operating power tools, re-generating electricity or even distilling water from water vapor—so as to subsequently produce hydrogen.

Compressed air and irrigation

Air tanks have small refrigerated compressors associated with them so that chilled air is 'spilled' out into the early morning prevailing breeze to condense water vapor into liquid water.

Refrigeration is an old and well-understood technology; further, many of the components associated with the air tanks would be standard parts.

Compressed air is driven through refrigeration unit(s) to substantially cool the air so that, as the chilled air is spilled out into the prevailing wind, the action has the effect of causing a substantial drop in air temperature. A drop in air temperature lowers the dew point of the air and causes water to condense.

The illustrations starting on page 33 show how sets of piezoelectric installations can, if both economic and climatic conditions are correct, be used to accelerate the condensation of liquid water from air. In this concept compressed air is used in combination with fog fences and trees to collect water for biomass irrigation.

Piezoelectric is the only active technology suggested, although electric fog fences have been patented that might provide an additional boost to harvesting moisture from the air. A discussion of electric fog fences, or other weather modification technologies is, however, beyond the scope of this work.

6

Active devices for biomass irrigation sites

The illustration below is an example of what a biomass growing area that is associated with a railroad right-of-way might look like. The site has three irrigation devices: Trees, such as Japanese Black Pine, are sited uphill from the rails and a gray stretch of fog fence can be see between the break in the trees.

The site also has a piezoelectric 'chilled air' system. The piezoelectric system works on compressed air. In the illustration two piezoelectric stations are located along the tracks and each compressed air station has air tanks. As trains roll over the piezoelectric crystals electrical energy is used to power air pumps that pressurized the tanks.

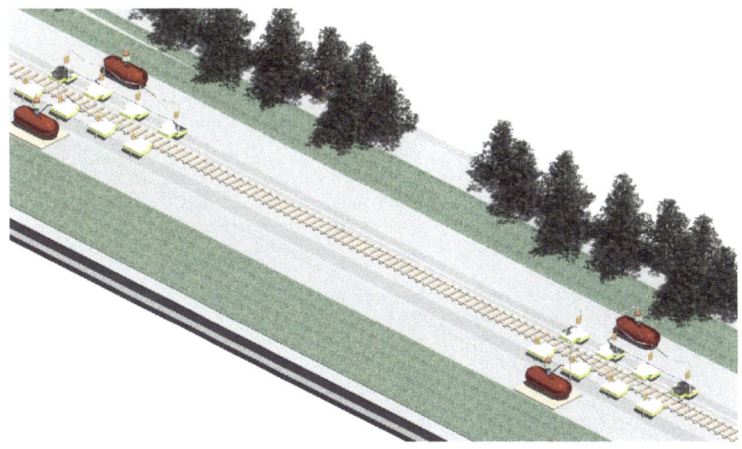

The illustration on page 34 is a closer view of one of the air tank installations. This is an example of how piezoelectric technology can be used to produce a limited electrical current and how that limited energy can be used to increase the efficiency of condensing water from air.

The following details how compressed air can be used in combination with other unconventional devices such as fog fences and trees to capture water from air.

Each air tank contains air under pressure. Depending on atmospheric conditions a valve opens that releases chilled air into the prevailing wind. The sudden drop in air temperature causes water vapor to condense.

The use of compressed air would be used only to reinforce or assist more passive means, such as trees or fog fences, when conditions indicate that (1) both trees and fences will not produce enough water and (2) when on-site sensors indicate that the release of chilled air will substantially increase the production of condensate.

On-site sensors use satellite rely to monitor air temperature, prevailing wind speed and direction and relative humidity. A central computer is programmed to actuate air chillers to release cold air into the prevailing wind when all of the conditions are favorable to the condensation of water. This type of active system takes much of the guesswork out of calculating and estimating how much water is needed to irrigate a given area of biomass.

7

Topography and irrigation

As best seen in the illustration on page 33 there are two biomass-growing areas. One growing area is uphill from the tracks and is associated with the trees. This is the primary (main) lineal biomass strip.

The other, secondary, biomass strip is on the other (downhill) side of the tracks. This growing strip is secondary, because it is at a distance from the trees and fog fence—the sources of irrigation water.

Railroad rights-of-way are typically built so that water drains to either side of the track. The rails are at a higher grade.

Trees on site should be place a sufficient distance away and uphill from the tracks to avoid flickering shadows on the tracks. One sure way to avoid shadows is to plant trees on the north side of the rails—In the northern hemisphere the sun arcs through the southern sky so shadows usually fall (in winter with a low sun) in a northerly direction.

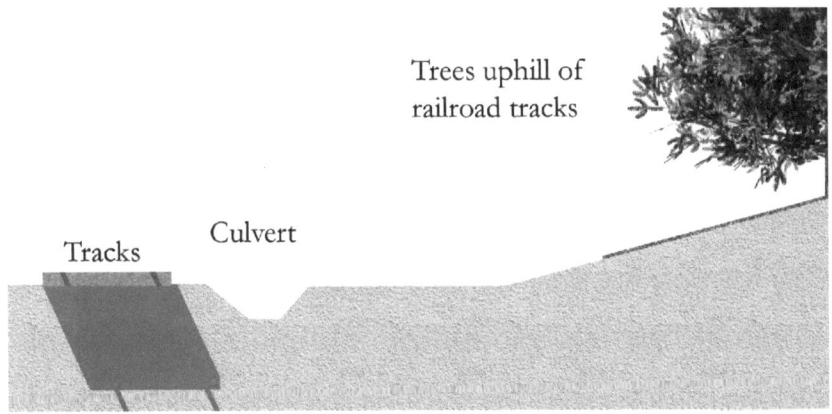

Trees uphill of
railroad tracks

Culvert

Tracks

Culvert(s) collect water draining off of a primary biomass area and distribute the water to the secondary growing area.   The fog fence(s) would be installed in gaps or breaks between trees, because the fences are made out of a mesh that can tear.

The uprights supporting the fence are plastic pipe that can be connected to pressurized air tanks so that, as air valves are opened, chilled air spills through the uprights into the prevailing breeze. Dropping the temperature of the air increases the amount of liquid condensation.

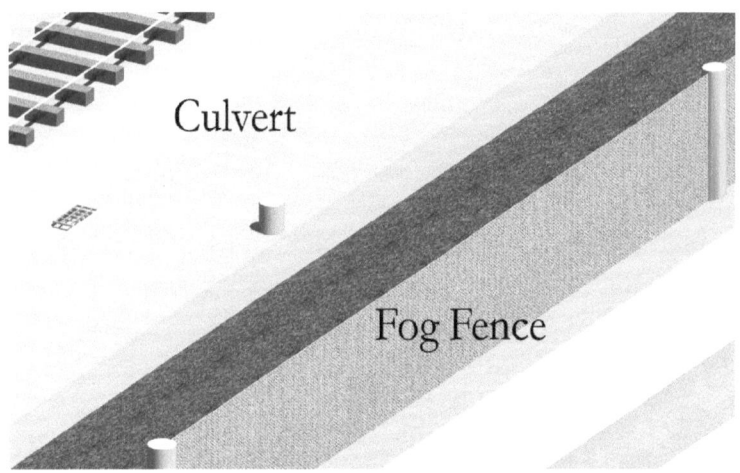

The use of piezoelectric technology has many potential applications in the design and maintenance of biomass growing sites.

Among those uses are chilling air to reinforce the use of passive devices such as fog fences and trees to condense water vapor.

Another use is to recharge the batteries of a sprinkler robot. The robot can be designed to be a simple mechanical device that rolls or 'trolls' up and down the culvert and throws accumulated water uphill onto the biomass crop.

A 'sprinkler robot' is not shown in the illustrations, because it is a simple machine for 'running up and down' the culverts and pumping water out of the culvert and irrigating surrounding biomass as the robot 'trolls' along the ditch.

8

Mechanized irrigation

There are four components in the following illustration:

1. A railroad track is seen in the upper left of the picture.

2. A piezoelectric/ compressed air tank is shown with a blue airline running from the tank to a standpipe with a valve and rainbird irrigator mounted on the standpipe.

3. A waterline (in red) is shown below the valve

4. A lateral airline (in blue) is seen running parallel with the waterline.

The railroad is necessary for two primary reasons:

❖ To power-up the PE compressed air tanks and

❖ To transport either biomass or a finished synthetic oil product.

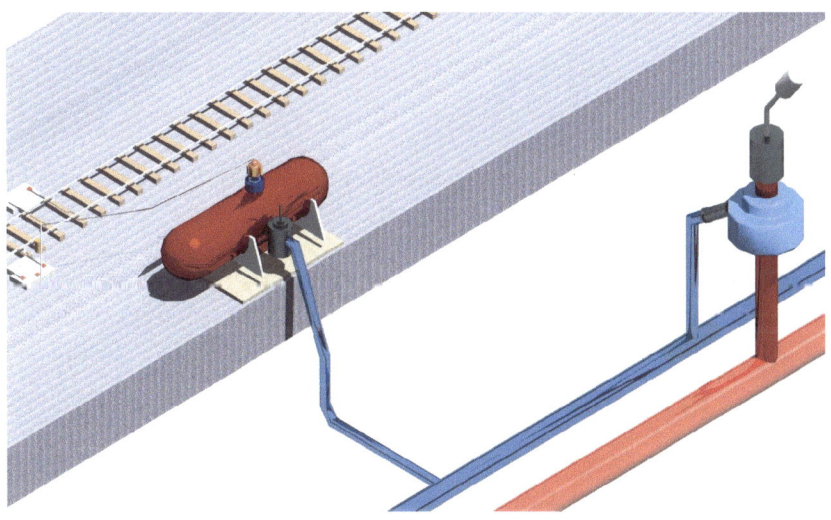

9

Urban irrigation

If a rail line runs through some urban areas water might be available at a cheap price. In this concept biomass is grown adjacent to railroad rights-of-way for the reasons stated, but also because the land directly associated with railroads tends to be completely devoid of any other use: It's commercial value is little and such land is, typically, littered, unkempt and ugly.

If railroads can be used to haul and mix a compost slurry composed of urban waste, paper, cardboard and other decomposable waste, the fewer miles the compost is hauled, the more profitable a biomass growing operation will be: Low input transportation charges translates into increase profitability if all other factors are equal.

The potential advantage of using urban areas (now depressed and rundown) is that city or urban water might be available at a cheap price.

While passive irrigation, discussed previously, centers on getting sufficient water to shrub land (desert) in remote arrears where drought is persistent, this variation is to use shrub urban land that is adjacent to abandoned factories and other urban infrastructure where the subterranean water utility infrastructure, now mostly abandoned with the factories and the railroad tracks still exists and is in working order.

In the illustrations a waterline is represented in red—that is city or county water that may have been installed a hundred years ago, but that is still a functional water delivery system; the biomass operation represents a type of urban farming enterprise and while the purchase of water represents an expense the land is cheap.

Nonetheless, water that is purchased from a utility must be considered to be expensive given the fact that the 'crop' is switchgrass.

The value of the biomass crop is determined by 'inputs' such as the cost of water, fertilizer and land. Another factor is that, given the age and condition, of some urban water utilities the water pressure in the lines may not be adequate for consistently irrigating large plots of land.

In this concept compressed air from the tanks is used to operate water pumps (air drives the pump turbine not electricity) that pull water out of the main and 'surcharge' the irrigation head so as to 'throw' the water over a much larger application radius.

Most conventional irrigation heads, such as Rainbird ® operate on the water pressure in the line. Typically, these devices adjust the radius of spray by turning or twisting an adjustment on the device itself, but in most instances the maximum radius of spray is about forty-five feet.

Depending on the site the biomass might be 'out of range' for conventional irrigation spray heads. By constructing PE compressed air tank facilities such as those illustrated compressed air is used to operate water pumps that increases the water pressure at the irrigation head. More pressure equals greater irrigation range.

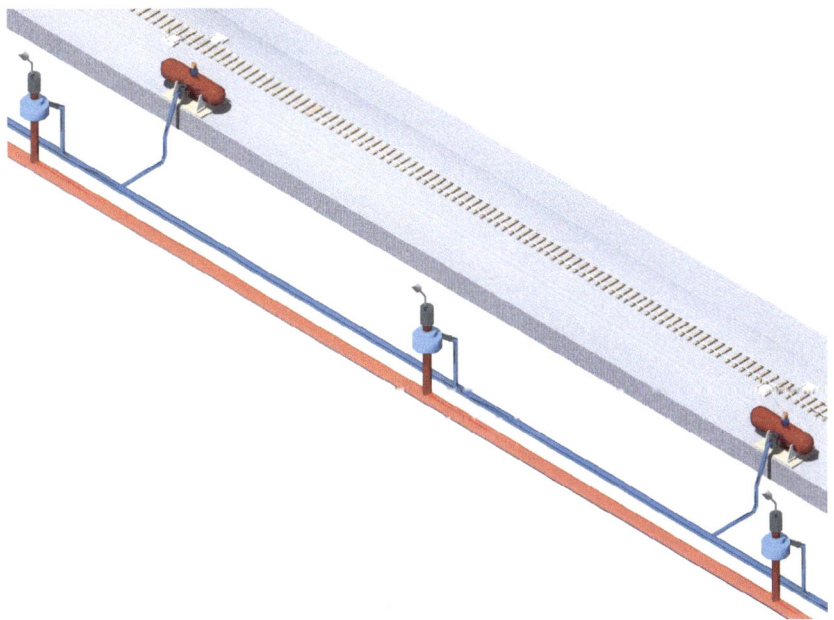

The objective is to design such biomass growing operations to be as cost-efficient as possible. To achieve greater economy the irrigation heads should also be detachable from the water standpipes:

1.   These installations would mostly be installed in urban areas that are depressed. So leaving expensive equipment such as a composite pump and irrigation heads on standpipes would be risking loosing expensive equipment due to theft.

2.   The compressed air tanks would be bolted to a concrete base and the tanks themselves would be heavy gauge steel so that it would require a crane or other heavy equipment to steal one.

By rotating the irrigation heads among different biomass growing areas it keeps the equipment in use, reduces irrigation expense for each growing area, secures the equipment (not left overnight). All of these factors function to increase both productive efficiency and economy.

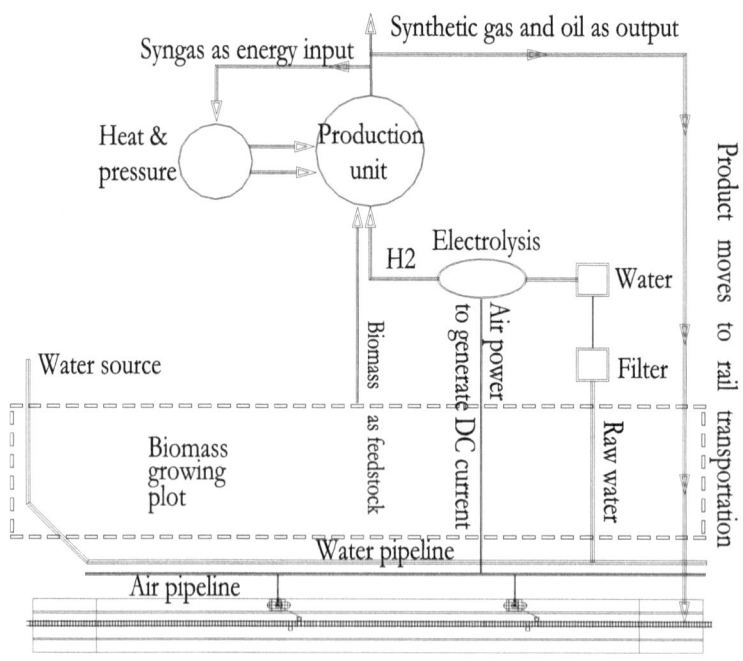

Introduction to Section 2

Expanded applications

Section 2 is about other potential applications of the biomass growing sites and makes, among other suggestions, that a 'redevelopment' of local manufacturing might be a good thing if such manufacturing can be integrated into a workable marketing plan:

❖ As with the economics of large-scale biomass production, the primary issue is cost. Can local manufacturing be developed that successfully competes with foreign competitors?

❖ The terrain, the soil or other environmental factors may not be suitable for biomass; or some small areas of the site, that might be 'at hand,' are not suitable for biomass production.

❖ Possibly growing trees or even shrimp or catfish. A site might have a water supply too small for conventional agriculture, but suitable for growing catfish.

❖ The basic concept is to find ways (other than tax dollars) to subsidize passenger train service.

The business objective is to generate additional revenues for the railroads:

❖ Take financing passenger train service out of the public budget; and

❖ Subsidize passenger train service by providing other commodities and services with the result that passenger train service is first-class and profitable.

Any and all means for providing profit should be explored. We are in a economic situation that does not allow us the luxury of turning our backs on a commodity or process that we deem 'beneath us,' because it involves shoveling dirt: We have squandered the cream of the earth's resources and now, faced with about seven billion hungry people, our attention must look down to the earth to see what opportunities we can scrounge.

We are entering an era where we will not have surplus energy, cheap energy based on plentiful petroleum. We can no longer avoid using ways to capture and effectively use less than optimal energy sources.

Part 2 of this section illustrates how other applications can be developed in combination with compressed air technology to provide a secondary, but economical means for co-generating energy and providing other commodities and services to enhance railroad profitability.

These are expanded applications of using piezoelectric salvaging of mechanical energy to co-generate electrical current and store that energy as compressed air. Section 2 takes the same technology, the same compressed air devices, and offers other applications aside from actively reinforcing the condensation of water from air.

Section 2

Expanded applications

Part 1

Do more than grow biomass

1

Background

This work suggests developing large-scale projects for the following reasons:

❖ Reduce our collective dependency on foreign oil.

❖ Develop a 'basket' of revenue producing activities to supplement and support passenger (and freight) train service.

If the United States, and other countries, are to reduce their dependency on oil and are also going to reduce our collective carbon footprint then one way to do both is to reestablish travel by train. If the history of passenger train service teaches anything, it is that:

1. Making profit from passenger train service is very difficult and

2. Some means of *subsidizing* passenger train service must be found independent of public subsidy.

3. And also, ways have to be found to foster competent and honest business management. (That might be the biggest challenge of all.)

If large-scale biomass production is undertaken then it opens the possibility that the railroads can meet the profit objectives of investors. No public money would be necessary.

Instrumental in the planning of these multi-functioning business plans is that 'unconventional' (non-fossil energy) should be developed and used.

One such technology, piezoelectric, has already been discussed and illustrated in combination with reinforcing and accelerating the condensation of water from air.

Technology such as piezoelectric has its uses, but also has its limitations in that the harvesting of mechanical energy from the passage of trails over the crystal modules is actually quite limited.

A fifty miles stretch of track, for example, would produce only about 5 Mega watts of electricity in an hour—that's enough 'juice' to compress air into tanks, to run at-grade rail crossing gates and recharge some batteries, but not much more.

Piezoelectric technology revisited

The same technology that is used to pump air can also be used to pump water or other fluids. Once a 'station' with piezoelectric installations is in place it is then a question of markets and economics as to what types of operations and products might be best developed at that location. This part looks at some potential marketing applications besides growing biomass.

2

Saltwater and hydroponics

Over fifty years ago Indian scientists experimented with using a combination of seawater (as an aqueous nutrient solution) and fresh water as a means to grow vegetables. Seawater has over ninety elemental nutrients so that when diluted with fresh water it can be used to grow many different varieties of vegetable and fruit crops.

Chemical tests are done to establish what nutrients are in the saltwater and how much fresh water needs to be added to an aqueous solution to provide a suitable growing solution for a specific plant crop. I suspect that one of the reasons that this idea has received only limited application is that seawater solutions need to be replenished and the 'spent' water needs to be discarded.

Saltwater forests

Trees such as Cypress flourish in saltwater conditions and shrimp and fish and other crops can be grown in deserts areas irrigated with saltwater. Saltwater forests provide carbon sequestration as well as allowing the 'spent' saltwater to filter through the desert.

❖ Growing shrimp and other saltwater stocks away from oceans isolates these stocks from native fish and other marine stocks—less cross contamination. Better control of marine diseases.

❖ Provide employment and another source of revenue for primary businesses and partners.

❖ Provide a source of wood and bamboo for the market.

❖ Japanese Black Pine and Austrian Pine are salt hardy species, but have been severely damaged by pests in the Northeast; growing hardy trees, combined with research for pest resistance, would be a huge market. Seawater forests offer great environmental benefits as well as revenues.

3

Water: irrigation and hydrogen

America and the world are running out of resources and not just oil. Water and oil are both essential resources. Throughout much of the southwestern United States drought conditions has persisted for years.

Air contains water vapor. Even desert air has about twenty percent humidity. Any rapid cooling of hot desert air results in condensation of vapor into liquid water.

Although a surprising amount of water can be condensed from warm (moisture laden) air it's not enough water to sell outright to ranchers, but enough water may be condensed from air to facilitate the production of hydrogen by disassociating water molecules into hydrogen and oxygen.

Hydrogen is also needed in the production of synthetic fuels and hydrogen is also a component of water. Water molecules can be disassociated by the passage of an electrical current in the presence of an electrolyte.

This is another potential application for using piezoelectric technology in remote locations to harvest both mechanical and thermal energy and to use that energy to further harvest water from air and to disassociate water into hydrogen and oxygen.

Any 'piezoelectric station' installed along the tracks that is part of a biomass operation can also be made integral to a synthetic fuels production depot—there is no point in hauling biomass miles to central chemical processing plant. It's just as easy to produce the synthetic fuel close to the biomass and then ship the finished product as fuel.

Hydrogen is essential to synthetic oil production. In this concept energy produced by piezoelectric technology or wind turbines, or micro hydropower is incorporated into an integrated energy infrastructure that stores energy either as compressed air or as hydrogen.

Solar and wind energy—if conditions are right

The illustration below is a view of a train on a siding with a grouping of wind turbines and a solar building. One advantage of developing a 'basket' of green energy is that green energy is not dependent on remote coal-fired power stations.

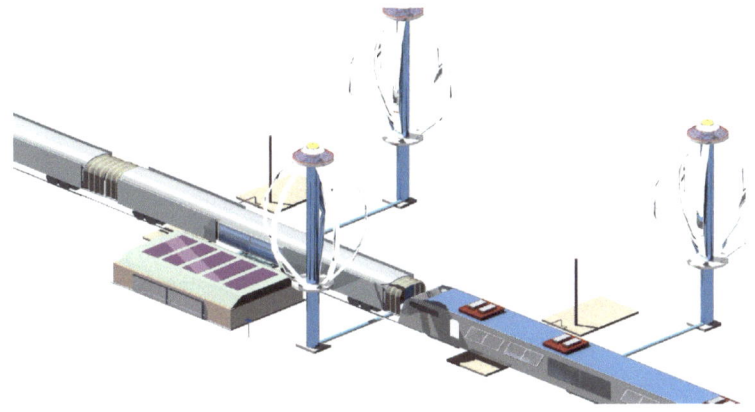

Agricultural refrigeration

One of the benefits of self-contained energy generation is the independent operation of refrigeration facilities at sites along the right-of-way. Solar buildings, in combination with energy stored as compressed air can provide a means for ranchers and farmers to store harvested produce without having to rely on motorized transportation that may or may not be reliably available in a post-petroleum economy.

We are not 'hedging our bet' by providing ranchers and farmers with alternative means of bringing their produce to market. Right now almost everything depends on automotive (truck) transport.

Composite energy infrastructures provide a means for providing electricity when the common grid may not be in operation or for providing irrigation or drinking water (on a limited basis) and for providing a means to refrigerate fruits, grains or other types of agricultural products while awaiting transportation by train.

Part 2

Manufacturing

1

Manufacturing as a systematic component

Can manufacturing be part of a green energy infrastructure in ways that create symbiotic benefits? Local manufacturing can reinforce and support other activities such as and energy co-generation and also be part of a multi-tiered energy infrastructure.

Green energy and manufacturing can be mutually beneficial in that manufactured goods need a distribution system and manufacturers need raw materials.

Manufacturing facilities also need heat, ventilation, power for tools and electricity. All of these products and services are possible with compressed air.

An underlying premise of this concept is that, due to many factors beyond the scope of this work, building and maintaining 'stand alone' transportation and energy infrastructures will be increasing difficult in the years ahead. We need to accomplish more with less—and we need to recycle much more than just bottles.

The United States has thousands of abandoned factories and many of these are adjacent to railroad tracks. As transportation by road becomes increasingly difficult associating manufacturing facilities near to railroads will make economic sense.

Manufacturing as a source of revenue:

1. If a manufacturing facility is located near the railroad right-of-way it can tap into a compressed air network for powering air tools, heating and ventilation and an economical means for electrical co-generation.

2. Manufacturing close to a railroad shipping center means less reliance on long-distance supply routes.

3. Locating manufacturing facilities as part of an energy/transit infrastructure makes possible greater economic self-reliance—many parts for the system's components, for example, can be manufactured locally. Less reliance on long-distance supply lines.

4. Geographically dispersing manufacturing provides maximum employment opportunities to local populations—whose goodwill is also necessary for a new transit/ energy infrastructure.

Paper and other waste products can be processed into electricity or otherwise used in the manufacture of other products by local manufacturing facilities.

Part 3

Railroads, the environment and the economy

1

Environmental issues

❖   The forests are long gone and

❖   The rats and tarantulas are not endangered.

The economic backbone of the concept is to use railroads, because (1) they already have a cleared right-of-way and (2) there is little likelihood of environmentalists raising hell, because there is little in the way of natural treasure to be ruined—trains already thunder along the tracks.

Fundamentally, there are three sets of problems and these problems are interrelated:

1. Corporate America must see profit in such projects.

2. Environmentalists must see minimal damage and, in fact, should see positive earth-friendly benefits of such projects.

3. The public, both national and local, should be well disposed toward these projects.

On the last issue, public acceptance, this type of projects offers specific project concepts that could provide real benefits for carbon sequestration and other environmental benefits while at the same time reducing dependency on petroleum, foreign and domestic.

Segments of the public tend to see things in the short term. In this regard this type of project would create jobs in both the rail transportation and in technology/ communications.

2

A stagnant economy and oil

The reason why railroads, and other corporate entities, might be inclined to 'go along' with the construction of a biomass growing operation is that this technology is central to creating and controlling a wide variety of moneymaking economic activities.

In a 'down' economy even railroads might be amendable to a little more income. This concept of using railroad rights-of-way is especially intended to create a total business environment as a means for achieving consistent profitability in an industry long known for its ability to lose money.

The American economy is stagnant. There are those that believe for a number of politically insensitive reasons that the years ahead do not bode well for the global economy:

1. Over-dependency on petroleum (and on cheap energy) has created an 'automotive' society that is simply unsustainable—even if people had jobs.

2. Cheap and plentiful petroleum made building tens of thousands of miles of Interstate highways possible, but the stagnation of the global and American economies renders this massive highway system largely unsustainable.

While figures lie and liars figure a rough comparison between highway infrastructure and rail infrastructure indicates that building and maintaining a two-lane concrete road costs about $15,000 per mile built and maintained.

One mile of rail (hypothetically) running along side a concrete road costs approximately $1500.00 per mile. A very rough estimate is that rail construction and maintenance cost is about ten (10) percent that of highways.

The challenge in designing and building these projects is that they must be 'well-along' before the public realizes that the golden age of the automobile is over, because if we actually wait for oil (expensive) oil to make its departure then we are left with no industrial transportation.

51

Again, from a business perspective, passenger railroads have a long history of criminally incompetent management and a God-given talent for loosing money. It is unlikely, however, that just good management can make passenger rail service pay. It is inherent in the nature of rail infrastructure that other sources of income be part of the business mix. This can be accomplished by designing *composite systems* that include biomass production.

A resource for local production

The redevelopment of passenger train service has benefits for saving energy and also for providing industrial transportation when automobiles running on seven dollar a gallon gasoline are shoved off of dilapidated overpasses onto roadways overgrown with weeds and potholes.

A revival of passenger train service would also be symptomatic that long-distance distribution of cargo had collapsed with the global market. In the coming decades the survival of many people will depend on their ability to live locally; and, a big part of that ability will be determined by making use of local resources.

If a project were built in the southwest United States where temperatures are hot in the summer, but where water is at a premium, then a project might be designed to use a combination of heat sources (including solar) to distill fresh water (hydrogen) from salt or brackish water.

What these projects have in common is the same underlying technology that supports a green initiative to produce synthetic fuels. Other processes and products depend on the local market.

The advantage of this type of localized energy technology is that the products or services that are produced are almost completely 'flexible' in that, for example, the technology may co-generate electricity—if that's what the local market needs—or it may output fresh water if that is what the local market needs. It may even produce a small quantity of electricity and a small quantity of water, but not both in large quantities, because there are limitations to green energy (piezoelectric/ solar and wind) inputs.

3

Utilizing sub-optimal resources

Many different services and projects are outlined in this work. These products and services range from distilling water to co-generating electricity to producing hydrogen for the subsequent production of synthetic fuels.

These services and products range from a means for local communities to be more energy efficient and more energy self-reliant—more independence from a more complex and therefore more vulnerable common grid—to a means for more independent (local) food production.

These products are brought together, bundled, into an encompassing technology that uses compressed air as a primary energy carrier. Air is contained in compressed air tanks as previously illustrated.

These activities range from co-generating electricity to distilling fresh water to growing saltwater shrimp. Activities brought under the control of an air technology, including local compressed air as a means to store energy.

Introduction to Section 3

Putting it all together

This section brings various components of the concept together to illustrate how synthetic oil can be produced locally while, concurrently, through the use of economical rail transportation, still have the benefits of *large-scale mass production*.

The concept is to develop very large biomass growing areas in out-of-the-way areas as well as in urban wastelands. In either case automation should be the dominant engineering tool used to standardize procedures, monitor growing areas, harvest biomass and either process the biomass into a final product or ship the biomass via rail to a production facility.

Associating biomass production with rail infrastructure provides a basis for synthetic oil production that has the benefits of economy of scale production even though the production is distributed over a large geographical area.

Here are the basic production components:

1. A railroad for transporting equipment and crews from one growing area to another; and, to provide compost and other inputs to each growing site; and, to ship biomass or synthetic oil;

2. Shrub land, either desert or abandoned urban sites as biomass growing sites;

3. Piezoelectric technology provides on-site electrical energy to compress air tanks or provide additional energy for irrigation and other auxiliary uses. (Such as generate hydrogen from water for use in synthetic gas/oil production.)

4. Equipment for irrigating, conditioning soil and irrigating the crop.

Section 3

Putting it all together

Part 1

Composite infrastructures

1

Economics of biomass production

Piezoelectric energy generation is expensive and inefficient so such technology will probably not find wide usage as a primary source of routine electrical energy; it's too expensive to run household appliances with it.

Piezoelectric technology, however, has the unique characteristic of generating useful amounts of electricity where the energy is needed provided a sufficient weight, such as the compression of rails by a train, are available.

Compressed air can be used to operate power tools, operate railroad rail crossings, recharge industrial batteries, operate water pumps, and even to drive small air turbines to co-generate a DC electrical current to disassociate water into hydrogen and oxygen as a component process in the manufacture of synthetic oil and synthetic gas.

Piezoelectric technology can also be rigged to provide 'chilled air' over a site, depending on prevailing wind, so as to enhance the condensation of moisture into liquid for a passive irrigation of biomass growing in areas of insufficient rainfall or affected with draught.

2

Conventional support infrastructure

History books teach us that Henry Ford 'put the world on wheels.' He put the world on wheels by reducing the price of his automobiles and he reduced the price by building large and efficient factories. The Henry Ford model of efficient production has more or less stayed with us over the years. Most chemical plants, to be efficient, are large operations:

❖ Electricity and water are brought into the plant,

❖ Workers and personnel are brought to the facility.

❖ Transportation in the form of rail or trucks is clustered around the plant and is organized so as to support plant operations.

❖ Raw materials are shipped to the plant.

In a sense the same organizational (business) principles apply here:

❖ The central operational modality is the biomass plot.

❖ Utilities are either onsite (Piezoelectric) or rail mobile.

❖ Operations are confined to the biomass plot and only marketable commodities such as synthetic fuels leave the site.

❖ Crews are brought into the site only for specific jobs and functions such as to install irrigate equipment where it is used and to oversee harvesting and biofuel production.

For *any* operation to be efficient economies of scale must be observed—one way or another. In this concept economies of scale are followed by minimizing the support infrastructure and by drastically reducing the ship cost of raw materials.

Consolidated operations

Every time a ton of material is moved one-mile energy is consumed and money is spent. Growing biomass for synthetic fuel production must be organized to be as efficient as possible both in the expenditure of energy inputs to the process and in money spent on materials and operations.

Parts 2 and 3 of Section 1 present a concept for recycling much of urban waste such as garbage, cardboard and plastics and processing these bulk commodities into compost to condition poor soils or for using these commodities directly in the production of synthetic fuels.

Urban waste materials, such as plastic bottles, have little intrinsic value and even in bulk quantities there is little value inherent in these materials—and, of course, that's why they are merely thrown away.

The odds of getting most cities (that are already financially broke) to pay more than they already are to dispose of these materials is very small.

Civil authorities, however, are likely to be very happy to assist in a process where those bulk materials were disposed of in a way that actually saves the city money.

Developing large-scale biomass growing areas requires some means for conditioning the soil, season after season, improving the soil. The best way to do that is to mix compost into the soil; and, urban waste (or most of it) can be processed into a compost mix using specially built railcars to haul and mix materials.

Cities collect fees from residents and businesses to collect the garbage and trash. Materials are then trucked to collect depots for some type of materials separation or simply trucked to landfills and dumped.

As indicated a lot of different materials ends up in the urban waste stream: poisons, paints and paint thinners, and thousands of other chemicals. The presence of these chemicals doesn't matter if the compost is used for growing biomass.

Shipping and economics of biomass production

Most Synfuel operations are limited by:

1.   The size and complicity of the operation (the plant)

2.   And, how far biomass must be transported.

This is one of the essential problems with getting maximum value out of biomass: It has to be found, processed into dry pellets or other components and then fed into the BTL plant. The problem is the processing of biomass and cost of shipping.

CTL plants or BTL plants are large, because that's how economy of scale is achieved. A throughput of a thousand tons of biomass is more efficient than a throughput of five hundred tons.

In this concept small, modular railroad cars, scalar production units are parked, as needed, on a siding after a biomass crop has been harvested, dried and processed into a component suitable for the BTL process.

The only component to actually leave the biomass site is the processed liquid (or gas) product(s). These are combined in rail tank cars and sent to market.

The rail cars that are mobile BTL processing units visit each biomass growing site once or twice a year and usually will travel along the mainline track from one field to the next as the growing seasons will be identical for each major growing area. So equipment is kept moving, kept productive, in this sequence. The irrigation equipment also only visits a growing plot as irrigation is actually needed.

Fundamentally, the concept is to bring the production facility, by rail, to the biomass site rather than harvest and ship a heavy, wet, low value commodity like switchgrass to a processing plant. The economy is in shipping liquid or gas fuel rather than shipping grass.

On large biomass growing sites a small railroad siding might be constructed so that a modular production car can leave the main track and park for the time necessary for pre-harvested, dried and processed biomass feedstock to be processed into intermediate or final products.

Other biomass sites may have a concrete platform built off the side of the tracks so that a modular production unit can be lifted off of a railcar and temporarily installed onsite.

In that case the railcar has a crane for lifting the production unit off of the flatbed car. The modular production unit is attached to the concrete pad and the railcar moves on.

A train may carry dozens of modular production units depending on the number and size of the biomass growing fields, although it would probably be established that agricultural practices are brought to bear so as to stagger and separate the harvesting times so that each single modular production unit is kept fully operational.

While each such production module(s) would be expensive using modern agricultural practices to regulate and control the timing of biomass harvests can reduce the number of such units.

The production goal, of course, is to bring the processing facilities to the biomass, but to do so in a way that minimizes equipment and transportation costs and maximizes profit.

Part 2

Creating value added infrastructures

1

Urban biomass

Anyone that has ever ridden on a train knows that trains travel through the underbelly of cities. The view from the window in any town is one of shacks, shanties, refuse and dilapidated and abandoned buildings.

The United States is a rich country, but you wouldn't know it traveling by rail through our cities. Rail rights-of-way often travel through the hearts of once-great urban areas. The real estate is still valuable, but we have to look beneath the litter and garbage to find a resource worth developing.

A previous section described a potential means for recycling much of the litter and refuse now dumped into landfills. Our present approach is again typical of a vast rich country willing to throw away resources that actually have energy value.

What is needed is a basic unifying infrastructure to make it economically feasible to salvage those resources and in the process convert those resources into marketable commodities like comfortable transportation and synthetic oil.

This work suggests that an underlying infrastructure needs to unify a multiplicity of different resources and bring value-added economics to scrub land and littered urban back ways. That unifying infrastructure should be to *use compressed air* in combination with railroads. Beautify (and prioritize) the railroads and beautify America at the same time.

2

Value added commodities

If the future belongs to sustainable energy then green energy projects must be engineered to be *consistently moneymaking*—otherwise the necessary investments will simply not be made.

Electricity within a *total energy infrastructure* is sold as a commodity such as synthetic oil—not as electricity. It's a total package. The electricity is repackaged, bundled into another product and sold as convenient comfortable travel or sold as liquid fuel. Value added.

Electricity can, if conditions are correct, also be sold simply as electricity. But, in the examples give below it's sold as a convenience to customers and so charging more for the electricity is seen as a convenience and not as an expensive necessity. Examples:

❖ Recharging vehicle batteries for commuters or for recharging buses or other equipment using battery power. (A service that the customer may or may not use.)

❖ For lighting and power in and around a *Passenger Transfer Site*—especially in the event of a common grid power failure. Safety, security, comfort. For a description of Passenger Transfer Site see '*Unconventional Transportation.*'

❖ Provide stored hydrogen to shopping centers for conversion into electricity using fuel cells.

❖ Provide stored hydrogen for conversion to electrical power (through fuel cells) for the local shopping center—to insure uninterrupted electrical supply. (Service, continuity of electrical supply)

❖ Provide hydrogen, water, and other elements for efficiently processing biomass into synthetic fuels.

Part 3

Suggested reading

Joyride to Infinity by
Dr. R.A. McConnell, PhD

Scott-Townsend Publishers, Washington, D.C.
A Scientific Study of the Doomsday Literature.

Beyond Growth
Dr. Herman E. Daly

Beacon Press. Boston
The economics of sustainable development

Power Down by
Richard Heinberg

New Society Publishers
Options and actions for a post-petroleum world

Living Within Limits by
Garrett Hardin

Oxford University Press
Ecology, Economics, and Population taboos.

Other writing by the same author:

Unconventional Transportation, 2nd edition

How buses and trains can be integrated into a single unified transportation system.

Unconventional Marine Carriers, 2nd edition

How 'quaint' modes of transportation such as water taxis can be designed to function within a total transit (green energy) system.

Novels:

The Sunburst Renegotiation

Hidden Jungle

Jacob Ebbtide

Shell Game

Earth, Mind and Murder

White Shadow

Becky's flight

www.ingramcontent.com/pod-product-compliance
Lightning Source LLC
Chambersburg PA
CBHW040838180526
45159CB00001B/232